Facing the Humanitarian Challenge

Towards a **Culture of Prevention**

Kofi Annan, Secretary-General
of the United Nations
September 1999

This text has also been
published as the Introduction
to the Secretary-General's
*Annual Report on the Work
of the Organization, 1999*

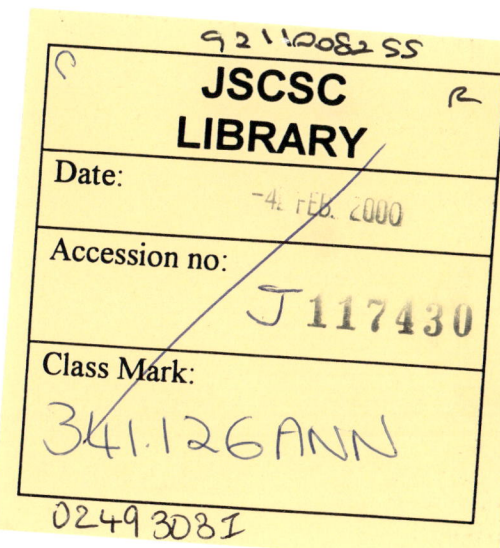

Published by the United Nations
Department of Public Information
New York, NY 10017
Copyright © United Nations, 1999
Facing the Humanitarian Challenge:
Towards a Culture of Prevention
ISBN 92-1-100825-5
Un...
Sa...
DI...

Introduction

Confronting the horrors of war and natural disasters, the United Nations has long argued that prevention is better than cure; that we must address root causes, not merely their symptoms. But aspiration has yet to be matched by effective action. As a consequence, the international community today confronts unprecedented humanitarian challenges.

The year 1998 was the worst on record for weather-related natural disasters. Floods and storms killed tens of thousands of people worldwide and displaced millions more. When the victims of earthquakes are included, some 50,000 lives were lost last year to natural disasters. Meanwhile, what had seemed a gradual but hopeful trend towards a world with fewer and less deadly wars may have halted. Armed conflicts broke out or re-erupted in Angola, Guinea-Bissau, Kashmir and Kosovo, and between Eritrea and Ethiopia. Other long-established wars, notably that in the Democratic Republic of the Congo, ground on largely unreported by the global media. Moreover, the impact of wars on civilians has worsened because internal wars, now the most

frequent type of armed conflict, typically take a heavier toll on civilians than inter-State wars, and because combatants increasingly have made targeting civilians a strategic objective. This brutal disregard for humanitarian norms—as well as the Geneva Conventions on the rules of war, the fiftieth anniversary of which we recently commemorated—also extends to treatment of humanitarian workers, who are all too frequently denied access to victims in conflict zones or are themselves attacked.

Confronted with renewed armed conflict and the rapidly escalating human and financial costs of natural disasters, our task is twofold. We must strengthen our capacity to bring relief to victims, but we must also devise more effective strategies to prevent emergencies from arising in the first place. The case for better and more cost-effective prevention strategies is my central theme in this report.

the Scope of the **Challenge**

The world has experienced three times as many great natural disasters in the 1990s as in the 1960s—while emergency aid funds have declined by 40 per cent in the past five years alone, according to the International Federation of Red Cross and Red Crescent Societies.

In the Caribbean, Hurricanes George and Mitch killed more than 13,000 in 1998, with Mitch being the deadliest Atlantic storm in 200 years. A much less publicized June cyclone in India caused damage comparable to Mitch and an estimated 10,000 deaths.

Major floods hit Bangladesh, India, Nepal and much of East Asia, with thousands killed. Two thirds of Bangladesh was inundated for months, making millions homeless. More than 3,000 died in China's catastrophic Yangtze flood, millions were displaced, and the financial cost is estimated to have been an astonishing $30 billion. Fires ravaged tens of thousands of square kilometres of forest in Brazil, Indonesia and Siberia, with devastating consequences for human health and local economies. In Afghanistan, earthquakes killed more than 9,000 people. In August this year, Turkey suffered one of the most devastating earthquakes in recent history.

In terms of violent conflicts, the most worrying development in 1998 was a significant increase in the number of wars. This is particularly troubling because the incidence and severity of global warfare had been declining since 1992—by a third or more, according to some researchers.

The humanitarian challenge is heightened by the fact that the international community does not respond in a consistent way to humanitarian emergencies. Media attention is part of the problem. The crisis in Kosovo, for example, received saturation coverage. The more protracted and deadly war between Eritrea and Ethiopia, and the resumption of Angola's savage civil war, received very little. Other wars went almost entirely unreported. Partly for that reason, responses to appeals for humanitarian and security assistance have been similarly skewed. It is my strong view that such assistance should not be allocated on the basis of media coverage, politics or geography. Its sole criterion should be human need.

I am particularly alarmed by the international community's poor response to the needs of victims of war and natural disasters in Africa. Where needs are pressing, if we are not true to our most basic principles of multilateralism and humanitarian ethics, we will be accused of inconsistency at best, hypocrisy at worst.

Worldwide Economic Losses
from Weather-related Natural Disasters

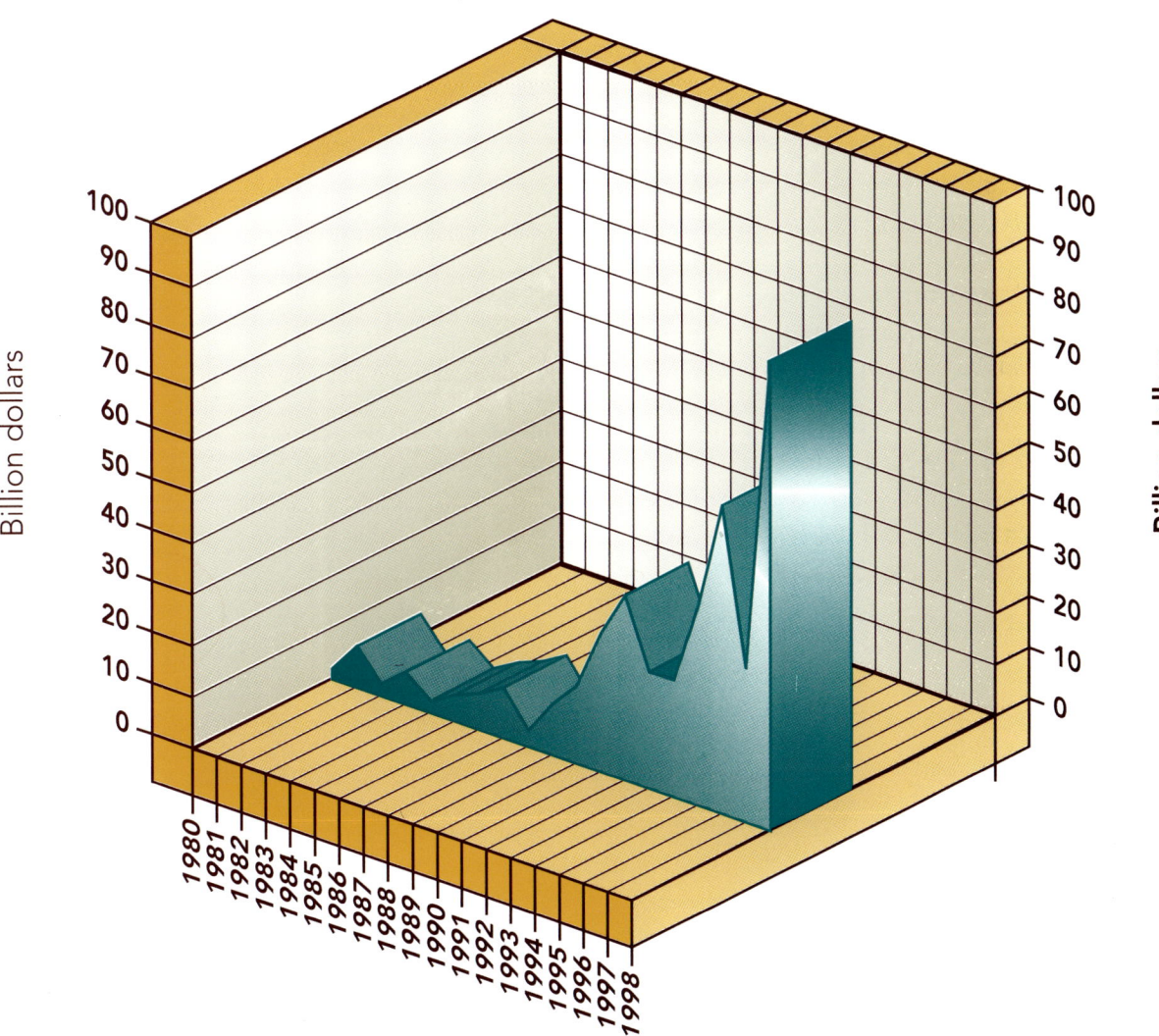

Source:
Worldwatch Institute (1999).

Understanding **Causes:**
the First Step to Successful Prevention

Devising preventive strategies that work requires that we first have a clear understanding of underlying causes. With respect to disasters the answers are relatively straightforward; war is a more complicated story.

Human communities will always face natural hazards—whether floods, droughts, storms or earthquakes. But today's disasters are sometimes man-made, and human action—or inaction—exacerbates virtually all of them. In fact, the term "natural" disaster has become an increasingly anachronistic misnomer. In reality, it is human behaviour that transforms natural hazards into what should be called *un*natural disasters.

Poverty and population pressures increase the costs of natural hazards because more and more people have been forced to live in harm's way—on flood plains, earthquake-prone zones and unstable hillsides. It is no accident that more than 90 per cent of all disaster victims worldwide live in developing countries.

Unsustainable development practices also contribute to the rising impact of natural hazards. Massive logging operations reduce the soil's ability to absorb heavy rainfall, making erosion and flooding more likely. The destruction of wetlands reduces the ability of the land to absorb heavy run-off, which in turn increases the risk of flooding. In 1998, an estimated 25 million people were driven off their lands into overcrowded and often disaster-prone cities by these and related forms of environmental malpractice.

Average Temperature at the Earth's Surface

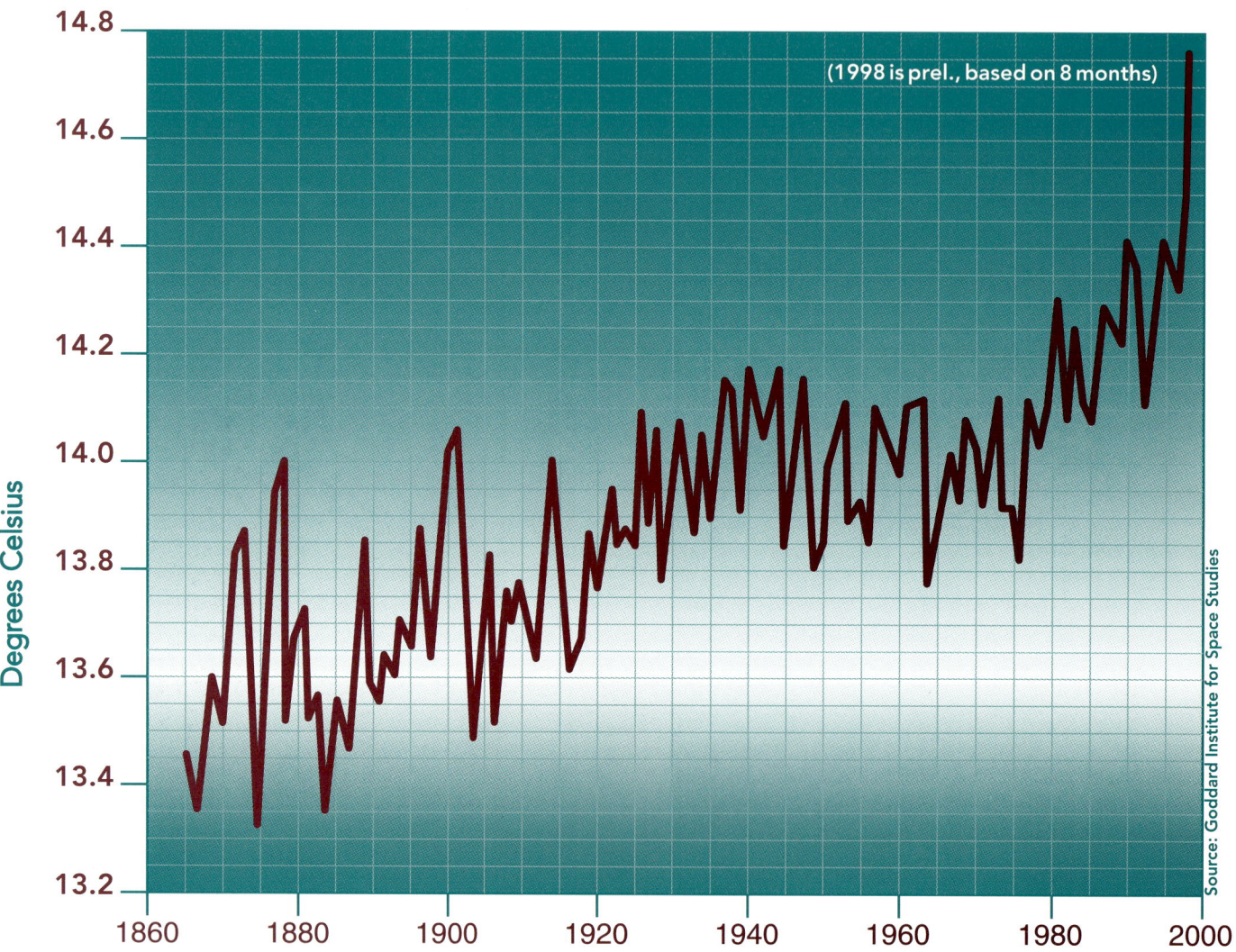

(1998 is prel., based on 8 months)

Degrees Celsius

Source: Goddard Institute for Space Studies

While the earth has always experienced natural cycles of warming and cooling, the 14 hottest years since measurements first began in the 1860s have occurred in the past two decades, and 1998 was the hottest year on record. Although still contested in some quarters, the evidence is steadily accumulating that the current wave of warming and the extreme climatic events associated with it are the product of increased carbon emissions, a large fraction of which is generated by human activity.

Understanding **Causes:**
the First Step to Successful Prevention

The causes of war are inherently more difficult to explain than those of natural events. Social behaviour is not subject to physical laws in the same way as cyclones or earthquakes; people make their own history, often violently and sometimes inexplicably. Thus, causality is complex and multidimensional, and it differs, often fundamentally, from war to war.

We can, however, identify some conditions that increase the probability of war. In recent years poor countries have been far more likely to be embroiled in armed conflicts than rich ones. But poverty per se appears not to be the decisive factor; most poor countries live in peace most of the time.

A study recently completed by the United Nations University shows that countries that are afflicted by war typically also suffer from inequality among domestic social groups. And it is this, rather than poverty, that seems to be the critical factor. The inequality may be based on ethnicity, religion, national identity or economic class, but it tends to be reflected in unequal access to political power that too often forecloses paths to peaceful change.

Economic decline is also strongly associated with violent conflict, not least because the politics of a shrinking economy are inherently more conflictual than those of economic growth. In some instances, the impact of radical economic reforms and structural adjustment programmes imposed without compensating social policies can undermine political stability. More

Understanding **Causes:**
the First Step to Successful Prevention

generally, weak governments—and, of course, so-called failed States—often simply lack the capacity to stop the eruption and spread of violence.

The shift from "war-proneness" to war itself can be triggered by the deliberate mobilization of grievances, and by ethnic, religious or nationalist myth mongering and the promotion of dehumanizing ideologies, all of them too often propagated by hate media. The widespread rise of what is sometimes called identity politics, coupled with the fact that fewer than 20 per cent of all States are ethnically homogeneous, means that political demagogues have little difficulty finding targets of opportunity and mobilizing support for chauvinist causes. The upsurge of "ethnic cleansing" in the 1990s provides stark evidence of the appalling human costs that this vicious exploitation of identity politics can generate.

But in other cases armed conflict has less to do with ethnic, national or other enmities than the struggle to control economic resources. The pursuit of diamonds, drugs, timber concessions and other valuable commodities drives a number of today's internal wars. In some countries the capacity of the State to extract resources from society and to allocate patronage to cronies or political allies is *the* prize to be fought over. In others, it is rebel groups and their backers who command most of the resources—and the patronage that goes with them.

Global **Armed Conflict, 1989-1998**

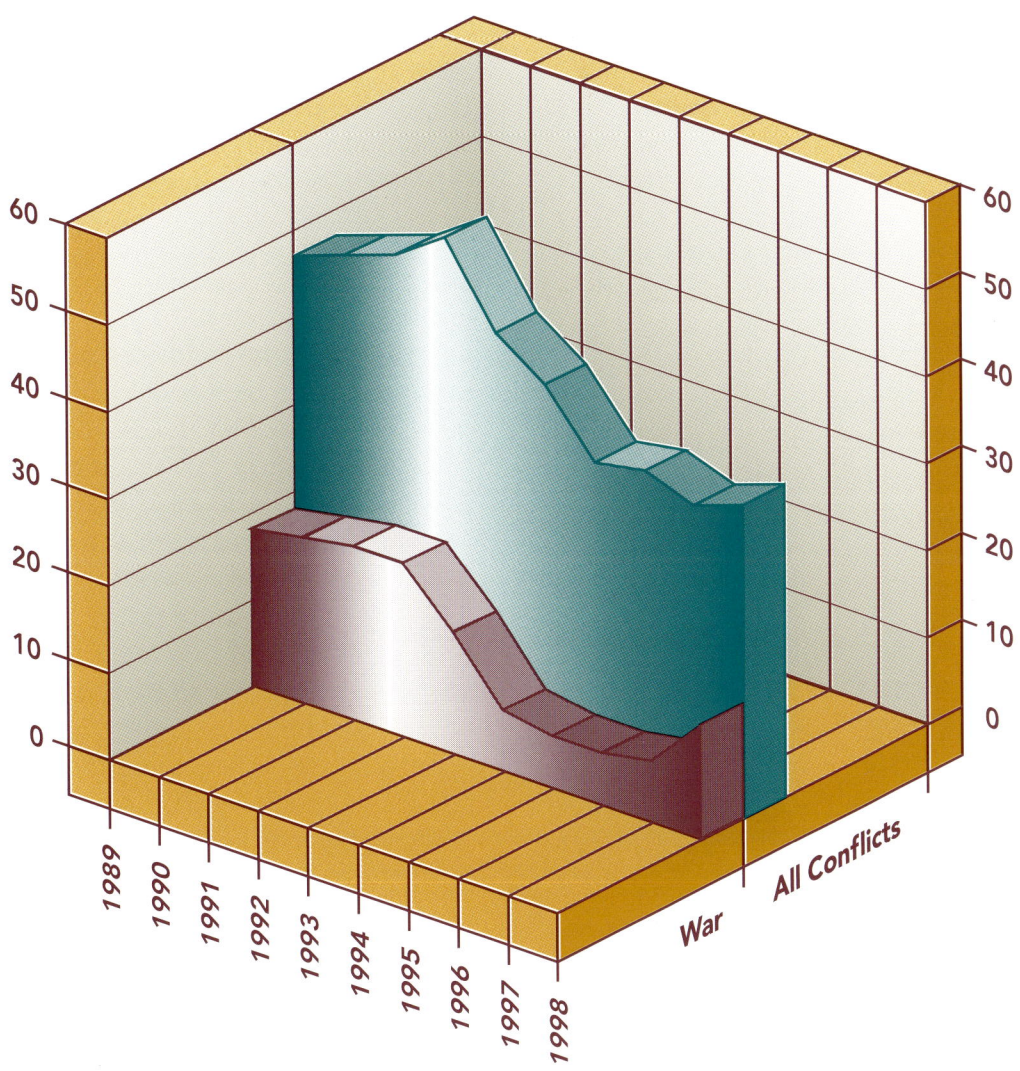

Source:
Wallensteen and Sollenberg,
University of Uppsala,
Sweden (1999).

Note:
"War" is defined as an armed conflict
in which there are 1,000 or more
battle-related deaths in a year.
The category "all conflicts" refers to
all armed conflicts with 25 or more
battle-related deaths in a year,
including wars.

Strategies for **Prevention**

Taking prevention more seriously will help ensure that there are fewer wars and less consequential disasters to cope with in the first place. There is a clear financial incentive for doing so. In the 1960s, natural disasters caused some $52 billion in damage; in the 1990s, the cost has already reached $479 billion. The costs of armed conflicts are equally sobering. The Carnegie Commission on Preventing Deadly Conflict estimates that the cost to the international community of the seven major wars in the 1990s, not including Kosovo, was $199 billion. This was *in addition* to the costs to the countries actually at war. The Carnegie researchers argued that most of these costs could have been saved if greater attention had been paid to prevention.

More effective prevention strategies would save not only tens of billions of dollars, but hundreds of thousands of lives as well. Funds currently spent on intervention and relief could be devoted to enhancing equitable and sustainable development instead, which would further reduce the risks of war and disaster.

Building a culture of prevention is not easy, however. While the costs of prevention have to be paid in the present, its benefits lie in the distant future. Moreover, the benefits are not tangible; they are the wars and disasters that do *not* happen. So we should not be surprised that preventive policies receive support that is more often rhetorical than substantive.

This is not all. History tells us that single-cause explanations of either war or natural disasters are invariably too simplistic. This means that no simple, all-embracing solutions are possible either. To address complex causes we need complex, interdisciplinary solutions. The fundamental point is that implementing prevention strategies—for wars or disasters—requires cooperation across a broad range of different agencies and departments.

Unfortunately, international and national bureaucracies have yet to eradicate the institutional barriers to building the cross-sector cooperation that is a prerequisite of successful prevention. For example, in national Governments as well as international agencies, departments that are responsible for security policy tend to have little knowledge of development and governance policies, while those responsible for the latter rarely think of them in security terms. Overcoming the barriers posed by organizational division requires dedicated leadership and a strong commitment to creating "horizontal" interdisciplinary policy networks that include our partners in international civil society.

Disaster **Prevention**

Disaster prevention seeks to reduce the vulnerability of societies to the effects of disasters, and also to address their man-made causes. Early warning is especially important for short-term prevention. Advance warning of famine facilitates relief operations; advance warning of storms and floods helps people move out of harm's way in time. Improvements in wide-area satellite surveillance technologies are revolutionizing the collection of early warning data relevant to disaster prevention.

United Nations agencies are playing an increasingly important early warning role. For example, the Food and Agriculture Organization of the United Nations provides vital warning on impending famines, while the World Meteorological Organization provides support for tropical cyclone forecasting and drought monitoring. The Internet is facilitating the real-time dissemination of satellite-derived and other warning data.

Greater efforts are also being made to improve contingency planning and other preparedness measures for disaster-prone countries, while major improvements in risk-assessment and loss-estimation methodologies have been identified through the International Decade for Natural Disaster Reduction (IDNDR). As a result of these and other innovations, national Governments are increasingly aware of the dangers and costs imposed by inappropriate land use and environmental practices.

There is also a growing consensus on what must be done. Stricter limits should be placed on residential and commercial development in hazardous areas—vulnerable flood plains, hillsides prone to slippage and earthquake fault zones. Construction codes should ensure more resilient buildings as well as infrastructure that can maintain essential services when disaster does strike. Such codes must, of course, be enforced. Sounder environmental practices are also needed, particularly with respect to deforestation of hillsides and the protection of wetlands. And because poverty rather than choice drives people to live in disaster-prone areas, to be truly effective, disaster prevention strategies should be integrated into overall development policies.

The experience of the IDNDR shows that a key to successful longer-term prevention strategies is broad-based cross-sectoral and interdisciplinary cooperation. The campaign to reduce carbon emissions and slow global warming illustrates what can be achieved with such cooperation. Working closely together and guided by the expert consensus that evolved in the United Nations Intergovernmental Panel on Climate Change, the scientific community and national and local governments, as well as non-governmental organizations (NGOs), have been highly successful in alerting the international community to the threats posed by global warming. Much remains to be done to transform this concern into effective action, however.

Here too we have ample evidence for the benefits of prevention. As severe as last year's floods in China were, the death toll would have been far higher without the extensive disaster-prevention efforts China has undertaken over the years. Floods on a similar scale in 1931 and 1954 claimed more than 140,000 and 33,000 lives, respectively—in contrast to 3,000 in 1998. Likewise, Hurricane Mitch claimed between 150 and 200 lives in one Honduran village, but none in an equally exposed village nearby, where a disaster-reduction pilot programme had been in operation for some time.

We should not underestimate the challenges, however. In some areas we still lack a broad scientific consensus on core issues and many questions remain unanswered. But the problem often lies not so much in achieving a consensus among scientists as in persuading Governments to resist pressures from vested interests opposed to change.

Resources are a pervasive concern. Some Governments, particularly in the poorest developing countries, simply lack the funds for major risk-reduction and disaster-prevention programmes. International assistance is critical here. And because

preparedness and prevention programmes can radically reduce the future need for humanitarian aid and reconstruction costs, such assistance is highly cost-effective.

Education is essential—and not just in schools. Many national Governments and local communities have long pursued appropriate and successful indigenous risk-reduction and risk-mitigation strategies. Finding ways to share that knowledge, and to couple it with the expertise of the scientific community and the practical experience of NGOs, should be encouraged.

For all of these reasons, it is essential that the pioneering work carried out during the International Decade for Natural Disaster Reduction be continued. In July of this year, the IDNDR Forum set out a strategy for the new millennium, "A Safer World in the 21st Century: Risk and Disaster Reduction". It has my full support.

Preventing War

For the United Nations, there is no higher goal, no deeper commitment and no greater ambition than preventing armed conflict. The main short- and medium-term strategies for preventing non-violent conflicts from escalating into war, and preventing earlier wars from erupting again, are preventive diplomacy, preventive deployment and preventive disarmament. "Post-conflict peace-building" is a broad policy approach that embraces all of these, as well as other initiatives. Longer-term prevention strategies address the root causes of armed conflict.

Whether it takes the form of mediation, conciliation or negotiation, preventive diplomacy is normally non-coercive, low-key and confidential in its approach. Its quiet achievements are mostly unheralded; indeed, it suffers from the irony that when it does succeed *nothing* happens. Sometimes the need for confidentiality means that success stories can never be told. As former United Nations Secretary-General U Thant once remarked, "The perfect good offices operation is one which is not heard of until it is successfully concluded or even never heard of at all." It is not surprising that preventive diplomacy is so often unappreciated by the public at large.

In some trouble spots, the mere presence of a skilled and trusted Special Representative of the Secretary-General can prevent the escalation of tensions; in others, more proactive engagement may be needed.

In September and October of last year, interventions by my Special Envoy to Afghanistan prevented escalating tensions between Iran and Afghanistan from erupting into war. This vital mission received little publicity, yet its cost was minimal and it succeeded in averting what could have been a massive loss of life.

Preventive diplomacy is not restricted to officials. Private individuals as well as national and international civil society organizations have played an increasingly active role in conflict prevention, management and resolution. So-called "citizen diplomacy" sometimes paves the way for subsequent official agreements. For example, former United States President Jimmy Carter's June 1994 visit to Pyongyang helped resolve a crisis over North Korea's nuclear weapons programme and set in motion a process that led directly to an agreement in October that year between the United States and the Democratic People's Republic of Korea. In the Middle East peace process, it was a small Norwegian research institute that played the critical initial role in paving the way for the 1993 Oslo Agreement.

In addressing volatile situations that could lead to violent confrontation, Governments are increasingly working in partnership with civil society organizations to defuse tensions and seek creative resolutions to what are often deep-seated problems. In Fiji,

for example, collaboration between NGOs and government officials, aided by the quiet diplomacy of some regional States, resulted in the promulgation of a new constitution and forestalled what many observers believed was a real possibility of violent conflict.

Early warning is also an essential component of preventive strategy and we have steadily improved our capacity to provide it, often in partnership with regional organizations, such as the Organization of African Unity. But the failures of the international community to intervene effectively in Rwanda and elsewhere were not due to a lack of warning. In the case of Rwanda, what was missing was the political willingness to use force in response to genocide. The key factors here were the reluctance of Member States to place their forces in harm's way in a conflict where no perceived vital interests were at stake, a concern over cost, and doubts—in the wake of Somalia—that intervention could succeed.

Complementing preventive diplomacy are preventive deployment and preventive disarmament. Like peacekeeping, preventive deployment is intended to provide a "thin blue line" to help contain conflicts by building confidence in areas of tension or between highly polarized communities. To date, the only specific instance of the

Preventing War

former has been the United Nations mission to the former Yugoslav Republic of Macedonia. Such deployments have been considered in other conflicts and remain an underutilized but potentially valuable preventive option.

Preventive disarmament seeks to reduce the number of small arms and light weapons in conflict-prone regions. In El Salvador, Mozambique and elsewhere, this has entailed demobilizing combat forces as well as collecting and destroying their weapons as part of the implementation of an overall peace agreement. Destroying yesterday's weapons prevents their being used in tomorrow's wars.

Preventive disarmament efforts are also increasingly directed towards slowing small arms and light weapons trafficking, the only weapons used in most of today's armed conflicts. These weapons do not cause wars, but they can dramatically increase both their lethality and their duration. I firmly support the various initiatives to curtail this lethal trade that are currently being pursued within the United Nations, at the regional level and by NGO coalitions.

What has come to be known as post-conflict peace-building is a major and relatively recent innovation in preventive strategy. During the 1990s, the United Nations developed a more holistic approach to implementing the comprehensive

peace agreements it negotiated. From Namibia to Guatemala, post-conflict peace-building has involved inter-agency teams working alongside NGOs and local citizens' groups to help provide emergency relief, demobilize combatants, clear mines, run elections, build impartial police forces and set in motion longer-term development efforts. The premise of this broad strategy is that human security, good governance, equitable development and respect for human rights are interdependent and mutually reinforcing.

Post-conflict peace-building is important not least because there are far more peace agreements to be implemented today than there were in the past. In fact, three times as many agreements have been signed in the 1990s as in the previous three decades. Some agreements have failed, often amid great publicity, but most have held.

Long-term prevention strategies, in addressing the root causes of conflict, seek to prevent destructive conflicts from arising in the first place. They embrace the same holistic approach to prevention that characterizes post-conflict peace-building. Their approach is reflected in the recent United Nations University study that found that inclusive government is the best guarantor

Preventing War

against internal violent conflicts. Inclusiveness requires that all the major groups in a society participate in its major institutions—government, administration, police and the military.

These conclusions are consistent with the so-called "democratic peace thesis", which states that democracies rarely go to war against each other, and that they have low levels of internal violence compared with non-democracies. The former proposition is still subject to lively debate among academic experts—in part over the changing meanings of "democracy" across time and geography. But the latter proposition is less controversial: in essence, democracy *is* a non-violent form of internal conflict management.

Long-term prevention embraces far too many strategies to be considered in detail in this essay. Here I will simply highlight three that are worthy of consideration, but have thus far received relatively little attention in the international community.

First, the international community should do more to encourage policies that enhance people-centred security in conflict-prone States. Equitable and sustainable development is a necessary condition for security, but minimum standards of security are also a precondition for development. Pursuing one in isolation from the other makes little sense. Security from organized violence is a priority concern of people everywhere, and ensuring democratic accountability and transparency in the security sector should receive greater support and encouragement from donor States and the international financial institutions. Moreover, since the overwhelming majority of today's armed conflicts take place within, not between, States, it makes good security sense in many cases to shift some of the resources that are allocated to expensive external defence programmes to relatively low-cost initiatives that enhance human—and hence national—security.

Second, greater effort should be put into ensuring that development policies do not exacerbate the risks of conflict—by increasing inequality between social groups, for example. In this context, the idea of "conflict impact assessments" should be explored further. Such assessments seek, via consultation with a broad range of stakeholders, to ensure that particular development or governance policies at the very least do not undermine security and at best enhance it. The model here is the well-established environmental impact assessment process, which accompanies major development and extractive industry projects in many countries.

Third, the changing realities of the global economy mean new challenges—and new opportunities. During the past decade development assistance has continued to decline while private capital flows to the developing world have risen significantly.

This has reduced the relative influence of donor States and international institutions in developing countries, while increasing the presence of international corporations. The private sector and security are linked in many ways, most obviously because thriving markets and human security go hand in hand. But global corporations can do more than simply endorse the virtues of the market. Their active support for better governance policies can help create environments in which both markets and human security flourish.

As must now be evident, the common thread running through almost all conflict-prevention policies is the need to pursue what we in the United Nations refer to as good governance. In practice, good governance involves promoting the rule of law, tolerance of minority and opposition groups, transparent political processes, a commitment to eradicate corruption, an independent judiciary, an impartial police force, a military that is strictly subject to civilian control, a free press and vibrant civil society institutions, as well as meaningful elections. Above all, good governance means respect for human rights.

We should not delude ourselves, however, into thinking that prevention is a panacea, or that even the best-resourced prevention policies will guarantee peace. Preventive strategy is predicated on the assumption of good faith, the belief that Governments will seek to place the welfare of the people

Preventing War

as a whole over narrow sectional interests. Sadly, we know that this is often not the case in practice. Indeed, many of the requirements of good governance that are central to prevention stand in stark contradiction to the survival strategies of some of the most conflict-prone Governments.

While providing incentives for progressive change can sometimes help, it is not something that the international community does well or often. The prospect of closer association with the European Union has served as a powerful tool for promoting tolerance and institutional reforms in several East and Central European countries, but few if any counterparts exist at the global level.

The fact that even the best-prepared prevention strategies can fail means that we can never completely escape the scourge of war. It follows that for the foreseeable future the international community must remain prepared to engage politically—and if necessary militarily—to contain, manage and ultimately resolve conflicts that have got out of hand. This will require a better-functioning collective security system than exists at the moment. It will require, above all, a greater willingness to intervene to prevent gross violations of human rights.

Demonstrable willingness to act in such circumstances should in turn serve the goal of prevention by enhancing deterrence. Even the most repressive leaders watch to

see what they can get away with, how far they can tear the fabric of human conscience before triggering an outraged external response. The more the international community succeeds in altering their destructive calculus, the more lives can be saved.

International security is, of course, the responsibility of the Security Council and responding to crises and emergencies will always be a major focus of Council activity. But, as Article 1 of the Charter reminds us, the Council is also charged with taking "effective collective measures for the *prevention* and removal of threats to the peace". Yet, reaction, not prevention, has been the dominant Council approach to dealing with conflict over the years.

Recently, however, the Council has shown increased interest in addressing prevention issues. This has been evident in the Council's extensive debate on post-conflict peace-building, and in its response to my report on the causes of conflict and the promotion of durable peace and sustainable development in Africa, which endorsed a range of conflict-prevention measures.

I greatly welcome these developments, and during the coming year I intend to continue the dialogue on prevention with Council members that started with the first Security Council Retreat, which I convened in June this year.

Conclusion

Today no one disputes that prevention is better, and cheaper, than reacting to crises after the fact. And yet our political and organizational cultures and practices remain oriented far more towards reaction than prevention. In the words of the ancient proverb, it is difficult to find money for medicine, but easy to find it for a coffin.

The transition from a culture of reaction to a culture of prevention will not be easy for the reasons I have outlined in this report, but the difficulty of our task does not make it any less imperative. War and natural disasters remain the major threats to the security of individuals and human communities worldwide. Our solemn duty to future generations is to reduce these threats. We know what needs to be done. What is now needed is the foresight and political will to do it.